Homo sapiens cyborgicus:
A Brief Essay

By Curtis R. Crim, BA CEO

ISBN: 978-0-9888255-4-3

Printed in the United States of America

First Printing

TABLE OF CONTENTS

Introduction

In this essay, I bring to attention the impact of modern day technological culture upon the DNA of modern human beings, also known as Homo sapiens sapiens.

In the past, changes in the environment have caused some traits of human beings to become vestigial or be radically modified. One example is the opposable big toe, which on human beings at some point in the past likely resembled that of the great apes.

Another example is the change in dentition, probably brought about by the use of fire to cook food. It is likely that humans once had a dentition that was more similar to that of a predator, as we see in chimpanzees. The use of fire to cook food, a form of technology, has caused a cultural influence that has reshaped the human jaw and teeth.

Modern day advanced digital technology will certainly have a *major impact* on future generations of human beings. The trend in what is happening to the human species can already be seen by observing how computer technology is changing the *behavior* of contemporary fully modern human beings.

These behavioral changes are compounded by the diet available to and consumed by the lower classes of society. One example is that the lack of foodstuffs containing adequate nutrition is causing damage to human DNA. The consumption of fast-food and junk food, which are almost totally lacking in any useful nutrition to the human physiology, is causing the devolution of future generations of humans.

My argument is that these technological, behavioral, and dietary changes in the way human beings experience life has lead to the emergence of new subspecies of Homo sapiens.

The classification of a subspecies of living organisms is sometimes determined by the culture or behavior of an organism, and not solely physical characteristics. In Chapter 3, I am using the Bili or Bondo apes of the Congo as an example to establish scientific precedent for this assertion.

Chapter 1: The Traditional View of DNA

For many years, the traditional view of DNA is that environmental pressures cause changes in the collective DNA in a population over time due to natural selection of the most beneficial traits. Darwin illustrated that these changes can take place in a population of animals in only a few generation, and over the course of only a few hundred years.

It is believed that the genes contained in human DNA dictate both physical characteristics, and also instinctive behaviors, such as the "fight or flight" reaction to life threatening situations.

The relationship between an animal and its DNA is seen as being static and one way. The organism is affected by the DNA, but the reverse is not seen as being true, at least not in the short term.

It is believed that it takes environmental pressures at least hundreds of years if not longer to cause a modification in a population's collective genes.

It is generally not believed that an individual can directly affect its own DNA simply through behavior and life experiences.

More recently, evidence of a more interactive system between a living organism and its DNA has been suggested.

Chapter 2: A New DNA Paradigm

Evidence has recently come to light that suggests a much more interactive systemic relationship between an individual and its DNA. It has been proposed that extreme experiences can have a direct and nearly instant effect upon the genes of an individual at the molecular level.

Example: The Rape-Suicide Victim: A rape-suicide victim is a person who has been raped, and reacted so severely on an emotional and psychological level that she (or he) cannot stand to cope with life, and as a result commits suicide. The memory of what was done to her is so terrible that she cannot stand to face life.

The examination of the DNA of these rape-suicide victims with an electron microscope has revealed distinct and measurable changes in the individual's DNA. This indicates that by putting individuals through emotionally unbearable events causes their genes to be affected immediately.

If the interpretation of the results of these experiments is accurate, then we can make a couple of leaps of logic that should be on course.

If a horrifying experience like rape can change an individual's DNA, then it is likely that torture would have a similar effect.

Is it reasonable to conclude that only terrifying and horrific experiences can have a short-term effect on DNA?

If an excessively painful experience can have a nearly instant effect, is it not possible that an over joyous experience can also have a measurable effect upon an individual's DNA?

The following leap of logic might appear unprecedented scientifically, but I believe that *every* experience in life not only maps itself onto the psychology of an individual, but it also has an impact upon its DNA on a subtle level.

I believe that the experiences an individual goes through in life not only make up his psychological profile, but they also dictate who he is on a physical level, and at the molecular level of the genes.

My theory is that human behavior (and therefore experience) is a reasonable basis for the classification of and taxonomy by which a subspecies is named.

Chapter 3: The Bili or "Bondo" Apes of the Congo

Bili (or Bondo) apes are classified as a subspecies of chimpanzee (*Pan troglodytes.*)

Although they are larger than their close cousins, Bili apes also demonstrate distinct cultural and behavioral differences.

For one thing, the Bili ape is known to the native humans of the region as "Lion killers". Western scientists have documented these apes predating upon large feline species, a behavior that is unique to their subspecies.

Bili apes also frequently make their bedding nests for sleeping at night on the ground, instead of in the trees as other chimpanzees do. This would appear to make them vulnerable at night, but it actually serves to keep them safer from humans and large feline predators, which have an easy time picking off chimps resting in the trees.

The Bili apes do not mate face to face as humans and orangutans do, but rather the male mounts the female from behind as do chimps, gorillas, and some other large primates such as baboons.

It has been observed that the Bili apes do not engage in cannibalism as other chimps and humans do. It has been argued that humans are more closely related to the Bili apes than to common chimpanzees because they are not cannibalistic.

Historically, this is completely inaccurate. Many instances and cultural traditions of human cannibalism have been documented including the eating of the children of Antioch and Jerusalem by Christian crusaders during the first crusade, the practice of cannibalism by native tribes of New Guinea, Jeffery Dahmer, and the Donner party of North America.

Because humans have always been cannibalistic, it is likely that we are more closely genetically related to the cannibalistic common chimpanzees than we are to the Bili apes that have not been observed to demonstrate this behavior.

Although the Bili apes have physical characteristics that distinguish them from other chimps and gorillas, for instance having larger heads and feet, and graying differently from gorillas, and having a sagittal crest like some gorillas but unlike chimps, it is the *behavioral* differences that seem to be the focus of physical anthropologists who have clarified their classification.

Unique behavioral patterns being emphasized as a basis for the taxonomical classification of a species (such as the Bili apes) lends scientific precedent and credibility to doing the same with human subspecies.

It is not simply by DNA and physical characteristics that newly emerging subspecies of humans should be classified, but by behavioral and cultural differences as well.

Chapter 4: Homo sapiens cyborgicus

The definition of the word, "cyborg" is "an organism that is composed of both biological and cybernetic parts." This is a very recent development in the experience of being human.

Although prosthetic limbs and body parts have been in use by human beings for thousands of years, going back at least as far as ancient Egyptian civilization, cybernetic body parts have only been available since the industrial revolution, and more specifically, the computer/digital technology revolution of the latter 20th century.

Although I have not heard it referred to as such, the last 40 to 50 years have been a technological revolution. I was born in the year 1963, and the world that I see now in 2013 is nothing like that which I saw back in the 1960's.

The invention and introduction into human culture of home computers, cellular phones, mobile computer technology, GPS technology, and the internet have completely revolutionized and changed the nature of human culture for all time. It is surprising to me that more people are not suffering from shock due to culture itself changing at too fast a pace.

The subject of this essay is the emergence of new species of Homo sapiens. Now that cybernetic human beings are a fact of modern culture, I find it relevant to address this issue.

Going back to my theory that everything that happens to an individual changes it on both a psychological *and* physical level, it follows that a human who has undergone such *radical* modification as having cybernetic technology integrated into its body could be viewed as a separate subspecies.

I now propose that the scientific community accept Homo sapiens with both cybernetic and biological components as the new subspecies *Homo sapiens cyborgicus.*

"Cyborgicus" is not a Latin term, so I am coining this term because there *is* no Latin translation, for obvious historical reasons.

I think that the word "cyborg" should be redefined. I believe that we should consider what exactly constitutes having cybernetic components. The common view would be that a cybernetic device has to be surgically implanted in an individual's body in order for that individual to be classified as being a cyborg.

I have observed the adoption of mobile computing technology by human beings in such a way that the behavior of individuals using this technology can be described as being radically different from individuals not equipped with said technology.

Further, I think that the interface between an organism and cybernetic technology should also be examined closely. If a surgically implemented connection to a cybernetic device can define an organism as being a cyborg, then a psychological connection or other form of dependence should also be considered a valid reason to classify an individual as being part of the human subspecies *cyborgicus*.

I have witnessed behaviors on the part of human beings using mobile cellular phone technology that can be described as nothing less than contemptible and socially irresponsible. I have witnessed human beings so distracted by and addicted to mobile technology that they become at best *partially* aware of the world around them. I have seen people attempting to operate cellular technology and automotive technology simultaneously, leading to *deadly* results.

Although they are technically classified as members of the human species, I only consider an individual a "person" if it behaves in a way that is responsible and legal. Even in states where it is not legal, cyborgs will continue to operate automotive vehicles without using hands-free cellular technology. Cyborgs become so egotistical and self-absorbed that they appear to care *nothing* for the lives they endanger while breaking the law.

If you own a cell phone, it might not be clear to you whether you are a cyborg. Most cyborgs have become dependant upon mobile technology without even becoming aware that they have become a member of a different subspecies.

I recommend the following experiment for the reader to self-determine whether it is a cyborg: First, pull out your cellular phone and destroy it with a hammer, then submerse the shattered remains in water. If you *literally cannot* do this, you are probably a member of the subspecies Homo sapiens cyborgicus.

If you are hesitant to destroy your cell phone, then pull the battery out and put your cellular phone in a drawer in your bedroom. Seal the drawer, and do not open it for a week. If you *can* do this, then you *might* be a member of the species of true modern humans, Homo sapiens sapiens.

I would also like to add that cyborgs are not the only newly emerging subspecies of humans currently on the planet Earth.

Chapter 5: Various Emerging Human Subspecies

In this chapter, I plan to define and discuss two other new subspecies of humans that are observed as existing on the Earth in 2013. I will also address in this essay the danger of conflict between subspecies of human beings, as this always leads to the elimination of all subspecies of Homo sapiens until only one remains.

I am proposing the following definitions of new subspecies of Homo sapiens:

Homo sapiens optimus (human aristocracy): I chose the Latin term "optimus" to designate the classification for this subspecies because it is the exact translation of the English term "aristocrat". Human aristocracy includes those who are billionaires. They are known as "illuminati", "The Bilderberg Group", and ICBM's (International Corrupt Billionaire Monsters). These people are traitors to all humankind, and are intent upon wiping out all other subspecies of human beings.

Homo sapiens servus (human slaves): I chose the Latin term "servus" to designate the classification for this subspecies of humans because it is the exact translation of the English term "slave". Human slaves own less than a billion dollars. Slaves are actually *encouraged* to "vote", but they get no real representation in their government. They are over taxed, victimized, and *literally* poisoned to death by the ruling subspecies of modern day human society.

In order to define these strata of society as different subspecies, it is critical to observe and compare their behaviors and cultures. I propose that when behavior and culture are examined we will be forced to conclude that these are not only different subspecies, but that the trend in the evolution in modern day culture will force additionally dramatic differences to develop between these two branches of humanity.

To do this, I will make a comparison of specific elements of culture, including education, politics, entertainment, and the obtaining of nutrition.

*** Education***

 The education available to Homo sapiens servus is vastly different from that which is available to the offspring of Homo sapiens optimus.

The aristocracy is able to afford the very best private schools, tutors, and teachers available in society. Their young are taught valuable skill-sets, given a serious education in multiple subjects, and are trained in how to control and manage slaves.

The slaves in society, on the other hand, receive virtually no education at all anymore. The education system of the USA was severely damaged by the illegal dictator George W. Bush, and was finished off by his slave, "president" Obama. When the US was *desperately* in need of the *doubling* of the annual education budget, it was instead curiously *halved* under Obama. This was an intentional act on the part of the subspecies Homo sapiens optimus. It is *unofficially* illegal to educate a slave in America in 2013. The United States of America now has the *lowest* standard of education in all of the modern industrialized countries. China is now at the *top* of the list in terms of educational standards.

The offspring of Homo sapiens servus are legally required to attend school, but the "public school system" is designed to indoctrinate slaves, and educate them only to the degree where they have the *minimum skills* needed to serve the ruling subspecies of humanity.

Further, history in these "school systems" is grossly distorted to the point where it is a crime. Children of the slave subspecies are not told the truth about the Crusades, the inquisitions, what was done to the Native Americans, or what happened during George W. Bush's illegal reign over the USA, among many other historical events.

The offspring of the slave subspecies of humans is lied to, brainwashed, miseducated, poisoned, and propagandized to rather than being educated.

I use the term "poisoned" because in "public schools" automated machines dispense beverages that contain substances that cause obesity, diabetes, damage to various organs such as the liver, give *no* relevant nutritional value, are narcotic and therefore addictive in nature, and cause *severe* brain damage. These beverages have been technically and intentionally designed and engineered to render the offspring of the human slave subspecies fat, diseased, distressing to observe visually, and mentally retarded.

The reasoning on the part of Homo sapiens optimus is that slaves with little to no education are easier to control, manipulate, and manage. They are also less threatening. The subspecies *optimus* is intentionally making the slave subspecies *servus* increasingly mentally retarded. They are retarding the slaves of society to make them more content complacent consumers for reasons of monetary profit.

*** Politics ***

The politics of the human subspecies *optimus* and those of the subspecies *servus* could not possibly be more different. They are exact polar opposites.

Homo sapiens optimus now controls the political systems of every country on the planet Earth. Illegally in most cases, they control all political systems using a shadow government installed within the legitimate government, which no longer retains any power. This subspecies pays no taxes, and deserves no representation; however, they retain all control over foreign and domestic policies in all countries, all of the time.

Homo sapiens servus, although over-taxed, no longer receives representation in any country on the planet Earth. They sometimes "vote" in elections which are designed to manipulate the slave subspecies into believing that they live in a "democratic" society, which is anything but the case. The slave subspecies of human beings has no hand in the making of the laws which are used to oppress them, and no say in the creation of foreign or domestic policy. Since the Patriot Act of Treason against the slaves of the USA, the slave subspecies also has no rights under the law. Further, the traitorous Bush administration has made it "legal" to capture and torture slaves to death without cause or purpose. They call this, "pre-interrogative processing."

*** Entertainment ***

Although all potential forms of entertainment are available to the ruling subspecies Homo sapiens optimus, one finds that they tend to prefer forms of entertainment that are not available to the slave subspecies.

Slaves on the other hand have many forms of distraction available to them, most riddled with forms of mental and psychological manipulation, brainwashing, lies, and propaganda. Complacent slaves have video games, the internet, movies, and "social" media available to them at all times. Most of these forms of entertainment are mental poison (like Facebook), and should be avoided by true humans (Homo sapiens sapiens) at all costs. One will note that cyborgs generally submerse themselves into these modern day forms of entertainment, oblivious as to their danger. However, cyborgs are automatically slaves from the viewpoint of the ruling subspecies, due to their only *partial* human existence. Most forms of modern day entertainment are designed to mentally retard the slave subspecies of humanity.

The ruling subspecies of billionaires, Homo sapiens optimus, on the other hand, enjoy sadistic forms of entertainment. They derive sexual satisfaction from causing wars that butcher the slave class in large numbers. They find it hysterical to destroy and manipulate legitimate forms of government, and feel that the slave subspecies is not *capable* of being self-governing. The billionaire subspecies actually experiences a *phobia* in relation to the concept of a truly democratic society.

When a member of the Homo sapiens optimus subspecies takes a commercial space flight (space itself as a domain was handed over to the ruling class under Obama when NASA was shut down), it has the option of using a satellite telescopic interface to select any slave on the planet and enjoy watching it be tortured and murdered. For a minor fee, it can also then have the selected individual served to him in a stroganoff or cacciatore. The ruling subspecies does not consider the life of an individual slave of any value.

*** Nutrition ***

The ruling billionaire subspecies of human beings has every form of nutrition available on the planet. Many have their own food production facilities and personal chefs. They have the option of eating fresh, delicious, and nutritious food at every meal.

In addition to that, they frequently enjoy consuming human flesh. Homo sapiens optimus regularly participates in the operation of farms that raise members of Homo sapiens servus for the consumption of their animal proteins. Most parts of these slaves raised for meat are used and consumed, including the blood, or sold back to fast foods such as McDonald's. It is believed that the traitorous

Bush family *alone* has consumed *hundreds* of human slave children.

The slave subspecies Homo sapiens servus on the other hand has only the *worst* sources of nutrition available. Being mentally retarded, most slaves eat junk food, fast food, soda pop, and highly refined carbohydrates laced with addictive narcotics such as High Fructose Corn Poison (HFCP). HFCP is found in almost all mass manufactured foods, including soda pop, bread, ice cream, chocolate syrup, ketchup, dressings, condiments, milk shakes, processed meats, prepared foods, and almost all fast food. It is not only addictive, but it destroys most human internal organs, and causes severe brain damage.

The ruling subspecies makes *vast* amounts of money selling this vile poison that is virtually free of any form of nutrition. They are able to sell highly nutritious foodstuffs, but they lack the motivation to do so.

The human body has hundreds of systems that are all required to function normally for an individual to enjoy good health. These many systems require only 20 nutrients to function correctly, so the lack of any *one* of those nutrients can have *devastating* consequences upon the health of the individual. The human body cannot remain disease-free without proper nutrition. Without eating properly, members of the slave subspecies make themselves and their families vulnerable to the possibility of contracting cancer or cardiovascular disease, among other detrimental conditions.

The reason that the ruling subspecies laces nutrition-free food with addictive narcotics and markets it to the slave subspecies is the same reason that mass-manufactured cigarettes are laced with highly carcinogenic substances: Members of the subspecies *Homo sapiens optimus* make *most* of their money on the pharmaceuticals industry.

The ICBM's are expected to make about a OVER FOUR TRILLION US DOLLARS on their pharmaceuticals trade alone in 2014. The pharmaceuticals industry is by far the most lucrative on Earth, totaling net profits that dwarf *all other industries* **combined**!

This is why they have always persecuted a brilliant healer named Burzynski. When the billionaires could not shut him down, they proceeded to steal his invention illegally.

The ruling billionaire subspecies Homo sapiens optimus also controls the media. They don't yet have total control of the internet, but you can see their brainwashing on many sites, and any site that displays Google Ads. Anything you see on Television, hear on the radio, or read in "news" papers or in magazines is anywhere from completely untrue to a spin on a very carefully selected portion of the truth. The **major** news networks are guiltier of these lies, this propaganda, and this brainwashing than any other form of media.

The media is used by billionaires to trick, manipulate, and brainwash the slave subspecies into thinking that it is okay to eat foodstuffs that are nutrition-free. This inevitably leads to illness and death, and members of the slave subspecies get to watch their family members die as the ruling subspecies of humans *laughs* all the way to the bank.

If you are a member of the slave subspecies Homo sapiens servus, then I suggest that you eat, drink, and feed your family only foodstuffs that you raise, brew, and produce yourself, rather than poisoning them on dangerous non-nutritional foods, and feel sorry for yourself as you watch your loved ones die.

*** Conclusions ***

Closely examining these major aspects of human behavior and culture, it is clear that the ruling class of billionaires is intentionally destroying the mental capacity of the slave subspecies of humanity. They are also poisoning them for the purposes of profiteering from the sale of medicine that is sold to slaves to cure the *symptoms* of their diseases, rather than curing the *cause* of their illnesses.

Another disturbing conclusion is that as the ruling billionaire subspecies makes themselves smarter and the slave subspecies increasingly mentally retarded and as they maintain their own health and make the slave subspecies complacent, lazy, obese, and diseased, they are also contributing to the further divergence between the two subspecies.

The ruling subspecies Homo sapiens optimus also raises members of the slave subspecies Homo sapiens servus as a source of nutrition. This in my opinion will lead to a situation that will inevitably

end in the elimination of one of these two
increasingly divergent subspecies of humanity.

Chapter 6: The Planned Obsolescence of Homo sapiens servus

The ruling billionaire subspecies of Homo sapiens optimus has planned the complete and total elimination of all members of the slave subspecies Homo sapiens servus, excepting a small population to be farmed as a source of animal nutrition.

The ruling subspecies Homo sapiens optimus among other things believes that they are Gods. They *actually think* that they are *better* than other contemporary subspecies of human beings. This, of course, is not the case. They actually are very similar to real human beings in most ways other than their culture and behavior. Even an aristocrat can bleed, and if so, it can also be killed. As long as they are mortal, they are nowhere near being Gods, although they have been being observed suffering from severe megalomania.

These International Corrupt Billionaire Monsters have enslaved you, destroyed your rights, and done away with democracy, justice, and representation of the slave subspecies in any and all forms of government.

The two groups of billionaires most responsible for the planning and demolition of the World Trade Center are the Bilderberg Group and Halliburton. Among those responsible is the close allegiance between the Bin Laden family and the Bush family in America. Families like the Bush's are guilty of *generations* of treason against the people and the now defunct Constitution of the United States of America. They are also guilty of war crimes, and treason against all of humanity.

The ruling class of billionaires has determined that the current infrastructure of society cannot continue to support such an increasingly large population of slaves, and has planned the obsolescence of the entire slave species of humanity (excepting their meat livestock, of course).

Fortunately for them, most cyborgs and slaves are so distracted, uneducated, unmotivated, and *outright retarded* that they are not even aware of their current situation. They are unaware that as time goes on, they become more divergent from the ruling subspecies physically, behaviorally, psychologically, and culturally, and will soon be obsolete.

Chapter 7: Summary and Conclusion

A long time ago, there was yet another subspecies of humans known as Homo sapiens neanderthalensis. They once coexisted with modern day humans, Homo sapiens sapiens.

This coexistence was a strained one, with both modern humans and Neanderthals competing for space and natural resources.

As a matter of history, the Neanderthals were completely wiped out. It is theorized that this came about due to war and direct conflict with modern humans, and because of being "pushed" out of the environment due to a lack of success in competing for needed resources. Some of their genes however, have been preserved due to absorption into the surviving modern human population. This can be seen by examining some members of the modern human population who live in southern Italy and Sicily, where the last remaining Neanderthals lived.

From this, it can be concluded that subspecies of humans have an inclination to conflict and compete with each other until only one remains.

The main competition will be between Homo sapiens optimus and Homo sapiens servus.

As members of the slave subspecies, should we not also be concerned about Homo sapiens cyborgicus?

Cyborgs are so heavily dependant upon technology that they cannot survive without it. Sadly, anyone who becomes *too* dependant upon cybernetic technology will inevitably die when that technology is no longer viable, supported, or available. The cyborgs are not really people, and I do not consider them among the living. Soon enough, they will all be dead, and the rest of us will be well rid of them.

The ultimate battle for survival will be between the ruling billionaires and the slaves. The billionaires already know this, and are working on the implementation of the elimination of a majority of the Homo sapiens servus subspecies. They do plan to keep their human meat farms in operation, because they enjoy slaves as a source of nutrition and culinary pleasure.

The ruling billionaire subspecies Homo sapiens optimus is immoral and monstrous. They plan to murder us all, because it is to their benefit to do so. The elimination of one of the two conflicting subspecies of humanity is inevitable.

If true humans are to survive, we must fight to destroy the billionaires before they destroy us. Coexistence between Homo sapiens optimus and Homo sapiens servus is not possible. Only one subspecies will survive. It's either them or us.

Appendix A: English to Latin Translations

English Terminology	Latin Translation
Slave	servus
Aristocrat	optimus
Law	lex or legis
Retarded	plumbeus or bardus
Cyborg	*no Latin equivalent*

Glossary

Homo sapiens sapiens: True modern day humans. Populations of this subspecies have become *very* rare. They are considered endangered and on the brink of extinction.

Homo sapiens optimus: The ruling billionaire class of society. This subspecies will inherently have the inclination to eliminate all other subspecies of humanity.

Homo sapiens servus: The slave class of society. This subspecies of humanity have no representation in their governments, are over-taxed, and are victimized and exploited by the ruling subspecies Homo sapiens optimus.

Homo sapiens cyborgicus: This subspecies of human beings are so physically, emotionally, psychologically, or behaviorally addicted or connected to cybernetic technology that they cannot survive without it.

Homo sapiens legis: Also known as *"pigs"*, this subspecies of human beings is used by the ruling subspecies Homo sapiens optimus to suppress, intimidate, exploit, victimize, murder, and control the human slave subspecies Homo sapiens servus.

Homo sapiens neanderthalensis: A now *extinct* subspecies of human beings who once inhabited Europe and Western and Central Asia.

Homo sapiens bardus: Commonly known as "the retarded", this subspecies of human being have through various genetic and cultural influences experienced the reduction of the functioning of higher cognitive processes to the point where they are nearly non-functional in society, and have no *discernable* intelligence or ability to learn.

Homo sapiens jerseyus: Human beings primarily inhabiting the East coast of Northern America in the New York/New Jersey region, this subspecies is primarily descended from Italian and Sicilian immigrants. They also possess some of the last remnants of Neanderthal genes in existence, and frequently appear strangely orange in skin tone. This subspecies demonstrates surprising behavior, and are easily excited. They are obsessed with egotism and glamour, and are preoccupied with physical violence and satisfying mating urges.

Homo sapiens deus: A human god. Two examples of which I am aware are Jesus Christ and Natalie Portman.

www.ingramcontent.com/pod-product-compliance
Lightning Source LLC
Chambersburg PA
CBHW060708280326
41933CB00012B/2345